Souad TAYANE

Organic Chemistry and Industrial Development

Souad TAYANE

Organic Chemistry and Industrial Development

Materials, Processes and Issues

ScienciaScripts

Imprint

Cover image: www.ingimage.com

This book is a translation from the original published under ISBN 978-620-2-26552-2.

Publisher:
Sciencia Scripts
is a trademark of
Dodo Books Indian Ocean Ltd. and OmniScriptum S.R.L publishing group

120 High Road, East Finchley, London, N2 9ED, United Kingdom
Str. Armeneasca 28/1, office 1, Chisinau MD-2012, Republic of Moldova, Europe
Managing Directors: Ieva Konstantinova, Victoria Ursu
info@omniscriptum.com

Printed at: see last page
ISBN: 978-620-8-40582-3

Table of contents

Chapter III: Transformation processes in chemistry

Chapter IV: Applications and Products of the Chemical Industry

Chapter V: Perspectives and Innovations in the Chemical Industry

Foreword

Industrial organic chemistry occupies an important place in modern society, providing essential materials and products for a wide range of sectors, from agriculture and medicine to the automotive industry and advanced technologies. This book aims to provide a comprehensive and up-to-date overview of this constantly evolving discipline.

Over the years, the processes and raw materials used in the chemical industry have been continually transformed to meet society's changing needs and growing concerns about sustainability. Today, the industry faces unprecedented challenges: it must not only maintain high production levels, but also integrate environmentally-friendly practices and develop technologies that minimize ecological impact. Advances in green chemistry, the circular economy and technological innovation are redefining traditional methods, paving the way for a more sustainable and ethical chemical industry.

This book is aimed at students, researchers, professionals and chemistry enthusiasts who wish to deepen their knowledge of industrial organic chemistry, explore its fundamentals and discover its innovations. Through clear explanations and concrete examples, it aims to provide a valuable tool for understanding how this industry is evolving to meet the needs of modern society.

General introduction

The organic chemical industry is at the heart of the global economy and touches almost every aspect of our daily lives. From plastics and pharmaceuticals to household products and textiles, industrial organic chemistry shapes our environment and provides solutions to technological, economic and environmental challenges. Its distinctive feature lies in its ability to transform natural resources such as oil, natural gas and even biomass into a multitude of finished products, each with a variety of applications.

This book begins by exploring the historical context of the chemical industry, providing an overview of its evolution from its origins to its integration into today's globalized economy. Raw materials, whether fossil or renewable, are examined in depth, as they form the basis of this industry and largely determine the processes and products derived from it. Each raw material requires specific transformation techniques, whether through steam cracking, polymerization or reforming processes, to produce finished products tailored to the requirements of different sectors.

Today's industry is facing increasing demands for sustainability. Environmental concerns are driving companies to incorporate eco-friendly practices, such as reducing CO_2 emissions, chemical recycling and the use of renewable raw materials. At the same time, the advent of green chemistry and nanotechnology is redefining the boundaries of what's possible, enabling the creation of materials that are more efficient, less polluting and sometimes even intelligent.

Finally, the development of artificial intelligence and molecular simulation technologies is opening up new prospects for innovation. These digital tools enable us not only to optimize production processes, but also to design materials with novel properties, opening the way to applications that would have been unimaginable just a few decades ago.

By bringing together all these facets, this book aims to provide an overview of industrial organic chemistry and its current challenges. Through its chapters, the reader is invited to understand not only the processes and applications of this industry, but also the challenges and transformations that are shaping it today. This technical and forward-looking journey opens a window onto the future of organic chemistry, in line with the demands of a society in search of sustainable progress.

Chapter I:

Background and trends in the organic chemicals industry

This chapter looks at the historical development of the organic chemical industry, from the earliest discoveries to its worldwide expansion. It examines the major producing countries and key players in the industry, as well as current trends, such as globalization and technological innovation, which are transforming the sector. By locating the industry's roots and developments, this chapter offers essential insight into its economic and technological importance.

I. Historical development and current context

I.1 Origins of the organic chemical industry

The roots of the organic chemical industry go back to the 19th century with the revolutionary discovery of Friedrich Wöhler, who synthesized urea in the laboratory, marking the birth of synthetic organic chemistry (1). The industrial production of coal tar-based dyes, such as aniline mauve by William Henry Perkin, paved the way for the industrial applications of organic chemistry (2). The rapid industrialization of chemical processes then led to the emergence of flagship products such as acetic acid, alcohols and phenol. The 20th century saw a second wave of development, with the advent of polymers (such as nylon and polyethylene) and pharmaceuticals, profoundly transforming everyday life.

I.2 Main producing countries and industrial players

On a global scale, the organic chemicals industry is dominated by industrialized countries such as the USA, Germany, China and Japan. The USA and Germany, for example, have a long history of chemical innovation supported by companies such as **BASF**, **Bayer**, and **Dow Chemical** (3). China has become a world leader, using its manufacturing base to become a production center (4). With a sophisticated industrial infrastructure, advanced research capabilities and favorable regulations, these countries represent poles of innovation, ensuring continuous production of new molecules and materials.

I.3 Changing trends in the context of globalization

Globalization has profoundly altered production dynamics in the chemical industry. Production facilities are moving to lower-cost regions such as Southeast Asia, enabling economies of scale while meeting growing global demand (6). However, pressure for more sustainable production is driving investment in green chemistry and the circular economy, particularly in Europe and North America, where regulations are stricter. Advanced technologies, such as artificial intelligence and biotechnology, are transforming research and production processes, favoring greener methods and optimal use of resources.

II. Economic importance of the organic chemical industry

II.1 Contribution to GDP in the main producing countries

In major economies, the chemical industry accounts for a substantial share of GDP. In Germany, it contributes around 5% of GDP, employing over 450,000 people (6). In the USA, the chemical industry generates a significant trade surplus and is one of the largest export sectors. In China, the industry supports rapid industrialization and contributes to the export of finished products, reinforcing the country's position as a global manufacturing center. The sector is also a major source of revenue for national infrastructure, development programs and research funding.

II.2 Role in the global supply chain

The organic chemicals industry occupies a central position in the global supply chain. It produces raw materials for a multitude of industries, from agriculture (pesticides, fertilizers) to pharmaceuticals (APIs - active pharmaceutical ingredients), consumer goods (plastics, textiles) and automotive (adhesives, coatings). For example, aromatic hydrocarbons such as benzene and toluene are essential raw materials in the manufacture of plastics, resins and detergents. The organic chemical industry acts as a multiplier, supporting other industrial sectors.

II.3. Global economic data

Figures provided by organizations such as ICCA and CEFIC show that the global chemical industry generates annual sales of $5,700 billion, creating direct and indirect employment for millions of workers (7). In developed economies such as the USA, it generates an annual trade surplus of around $28 billion, while in Europe, it is a driver of technological innovation, investing in research projects such as REACH for safety and sustainability standards. In Asia, the rapid growth of China and India is helping to boost the competitiveness and efficiency of global chemical production.

III. Industry structure and organization

III.1 Definition of the main sub-sectors

III.1.1. Basic chemistry

This sub-sector encompasses the production of simple organic compounds, such as ethylene, methanol and ammonia, used as building blocks for many other chemicals and materials. Basic chemicals depend on large industrial complexes and aim for significant economies of scale to produce at low cost.

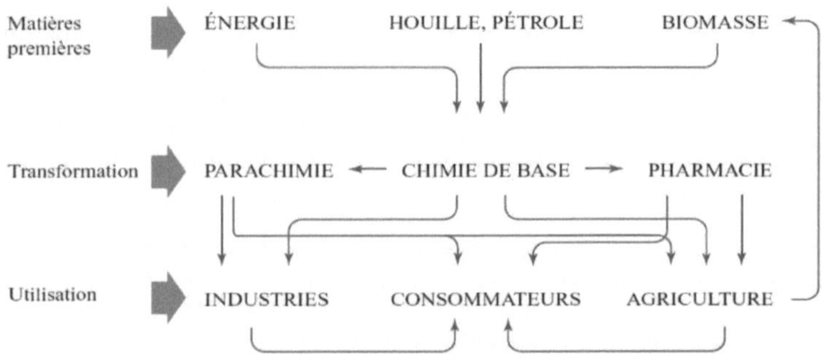

Figure 1: Industrial and economic activity in chemistry.

Between the natural sources of raw materials and their uses, industrial activity itself occupies an intermediate position. Basic chemistry" produces compounds not found in nature but with no direct applications, from which "parachemistry" and "pharmaceuticals" prepare products for end-users.

III.1.2. Fine chemistry

Producing complex and often customized molecules, fine chemicals include intermediates for specific applications, such as pharmaceuticals and agrochemicals. This sub-sector is characterized by sophisticated manufacturing processes, strict quality control and high purity.

III.1.3. Parachemistry

This field covers everyday consumer products such as detergents, cosmetics, paints and inks. The sector is influenced by consumer demand and must adapt to safety and sustainability standards. The parachemistry sector is particularly dynamic, constantly innovating to meet new expectations in terms of health and respect for the environment.

III.1.4. Pharmacy

This sector is dedicated to the production of medicines and healthcare products, often in collaboration with the fine chemicals sector for active substances. Pharmaceuticals are highly regulated, requiring particularly strict safety and efficacy standards, with production norms that guarantee the quality of medical treatments.

III.2 Interconnections between these sub-sectors

The interconnection between these sub-sectors is essential for industrial synergy. For example, basic chemistry supplies elements such as propylene, which is then used in fine chemistry to create high-purity intermediates needed in pharmaceuticals or parachemistry. Advances in one sub-sector, such as innovations in fine chemicals, influence the products developed in other sub-sectors, contributing to complex value chains that promote the flexibility and resilience of the chemical industry (8).

IV. Challenges and opportunities for the organic chemicals industry

IV.1 Environmental and regulatory challenges

The organic chemicals sector is facing major challenges linked to environmental regulations and sustainability requirements. Regulations, notably in Europe with the REACH (Registration, Evaluation, Authorization, and Restriction of Chemicals) regulation, require companies to minimize risks to health and the environment. These constraints encourage companies to develop less-polluting processes and more environmentally-friendly products, but require substantial investment.

IV.2 Technological innovation and digital transformation

Digital transformation is redefining production and management methods in the organic chemicals industry. The integration of artificial intelligence, connected sensors and big data platforms is enabling process optimization, cost reduction and improved quality control. In addition, biotechnologies such as synthetic biology and enzymatic catalysis enable the creation of innovative products while reducing the ecological footprint, opening up possibilities for green chemistry and bioplastics.

IV.3. Growth prospects and new markets

Growth prospects are strong in sectors linked to green chemistry, such as the production of biodegradable materials, biofuels and biosourced plastics. New emerging regions, such as Africa and Latin America, present expanding markets for chemical products, while offering opportunities for international collaboration and sustainable development.

References

1. Brock, W. H. (1993). *The Fontana History of Chemistry*. Fontana Press.
2. Travis, A. S. (1993). *The Rainbow Makers: The Origins of the Synthetic Dyestuffs Industry in Western Europe*. Lehigh University Press.
3. Spitz, P. H. (2003). *Petrochemicals: The Rise of an Industry*. Wiley.
4. Ziegler, R. (2001). *The Story of BASF in China*. BASF Press.
5. OECD (2018). *The Future of Petrochemicals*. OECD Publishing.
6. Verband der Chemischen Industrie (2019). *The Chemical Industry in Germany*.
7. CEFIC (2021). *Facts and Figures of the European Chemical Industry*.
8. Matar, S., & Hatch, L. F. (2000). *Chemicals: Essential Chemicals for Industrial Processes*. Gulf Professional Publishing.

Chapter II:

Raw materials and sources for chemistry

This chapter explores the sources of raw materials for organic chemistry, including oil, natural gas, coal and biomass. Each section details the extraction and transformation processes of these resources, as well as their specific uses in industry. The chapter highlights the emergence of renewable raw materials in the context of sustainable development, comparing their advantages, disadvantages and future prospects.

I. Oil and natural gas

Petroleum was initially used as a fuel for lighting, and the first attempts at refining were aimed at improving this application, as crude oil burns poorly (production of "kerosene" or lamp oil, 1850) (1). It then appeared as a source of calorific and mechanical energy (first for heating steam boilers, then for internal combustion engines), and it was in 1940 that oil began to be systematically exploited as a source of organic raw materials, constituting the field of petrochemistry (2). Global crude oil production grew rapidly, supported by increases in countries such as the USA, Brazil and Iran, despite OPEC production cuts (3). At the same time, crude oil production grew extremely rapidly:

Table 1: Annual world crude oil production

Year	**1860**	**1900**	**1950**	**1976**	**1993**	**2001**	**2007**	**2015**	**2020**	**2023**
Production (million tonnes)	0,07	20	524	2 918	2 969	4 450	3 603	4 389	3 789	4 216

In 2023, production rose slightly by 1% on the previous year, to around 4,216 million tonnes. Overall production was supported by increases in countries such as the USA, Brazil and Iran, despite production cuts by OPEC (Organization of the Petroleum Exporting Countries) to stabilize the world market (3).

I.1 Composition of oil and natural gas

Petroleum is a complex mixture of hydrocarbons that may include alkanes, cycloalkanes, aromatics and, to a lesser extent, heteroatoms such as sulfur, nitrogen and oxygen. Its composition varies according to its geographical source: light oils, for example, contain a high proportion of low-molecular-weight hydrocarbons, while heavy oils are richer in long-chain components. This diversity is important for the chemical industry, as each type of oil can be adapted to specific uses, depending on the properties of the fractions obtained during refining (4).

Natural gas, meanwhile, is mainly composed of methane (CH_4), accounting for around 70-90% of its content, with varying proportions of ethane, propane, butane and other light hydrocarbons. It is often combined with non-hydrocarbon components such as carbon dioxide (CO_2), hydrogen sulfide (H_2S) and dinitrogen (N_2). This composition makes natural gas a valuable raw material not only for heat and power generation, but also for the synthesis of chemical compounds such as hydrogen, ammonia and methanol (5).

I.2 Oil refining processes: distillation, cracking, reforming, steam cracking

Crude oil is treated by a series of processes that transform it into lighter products and compounds that can be used in a variety of applications. These processes include :

I.2.1. Distillation

Atmospheric distillation is the first step in refining, where oil is heated to separate its components according to their boiling point. The fractions obtained include gases, gasolines, naphthas, kerosene, gas oils and heavy residues. This operation is essential to isolate the basic components for further processing (6).

I.2.2. Cracking

In catalytic cracking, heavy hydrocarbons are broken down into smaller, lighter molecules, often in the presence of catalysts such as zeolite. This process converts heavy fractions, such as gas oils, into gasolines and light alkenes (such as ethylene and propylene), which are essential for the production of plastics and synthetic rubber (7).

I.2.3. Reforming

Catalytic reforming converts light naphtha into aromatics and hydrogen, in the presence of catalysts such as platinum. Aromatics, such as benzene, toluene and xylene, are precursors for many industrial chemicals, while hydrogen is used in various industrial processes and for refining itself (8).

I.2.4. Steam cracking

The steam cracking operation has now moved away from the fuel sector to focus on the production of alkenes, such as ethylene, propylene, butenes, butadiene and isoprene (or 2-methylbuta-1,3-diene), and, to a lesser extent, aromatic hydrocarbons such as benzene, toluene and xylenes. These molecules, essential for subsequent synthesis in the chemical industry, are obtained in relatively small quantities during catalytic cracking, hence the growing importance of steam cracking (9).

This process involves exposing light hydrocarbons, such as naphtha and diesel, to extremely high temperatures (around 800°C) for a very short period of time (a fraction of a second). These hydrocarbons are first vaporized and mixed with steam to promote the reaction. A typical steam cracking unit can process over a million tonnes of hydrocarbons a year, with annual yields reaching around 300,000 tonnes of ethylene, 50,000 tonnes of butadiene and 90,000 tonnes of benzene. Annual ethylene production in France using this process is close to 3,000,000 tonnes, underlining its strategic role in the supply of this key molecule for the chemical industry.

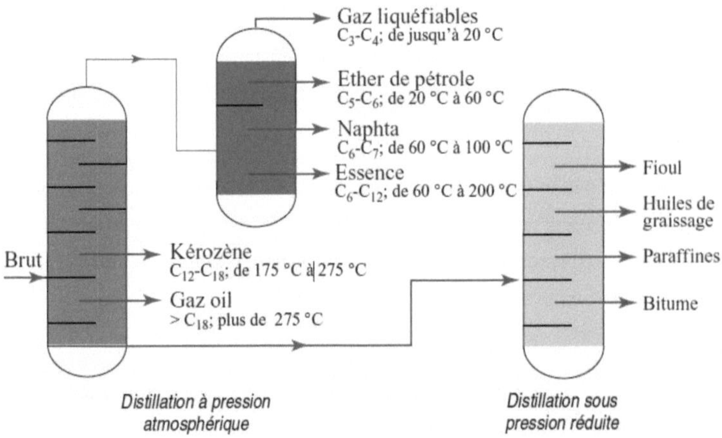

Figure 2: Crude oil distillation

Due to the very wide temperature range covered by the boiling points of petroleum constituents (from 0°C to several hundred degrees), it is not possible to carry out all the necessary fractionations in a single operation. After "roughing" in a first column, the lighter fractions are separated in another installation. The heavier fractions are redistilled under reduced pressure: this lowers their boiling point and prevents their decomposition by heat.

Table 2: Oil consumption by continent in 2023, expressed in millions of barrels per day ( )3

Continent	Daily consumption (millions of barrels)
Asia-Pacific	38,2
North America	23,8
Europe	10,5
Latin America	6,8
Africa	4,2
Middle East	9,5

I.3 Importance of hydrocarbons in the chemical industry

Hydrocarbons derived from oil and natural gas are fundamental to the production of many synthetic materials and chemicals. Ethylene and propylene are basic raw materials for polymers such as polyethylene and polypropylene, used in applications ranging from packaging to textiles. Aromatic hydrocarbons, obtained by reforming, are precursors for solvents, resins, detergents and pharmaceuticals. In short, hydrocarbons form the backbone of the modern chemical industry and are essential to many industrial sectors (10).

II. Coal and Carbochemicals

Hydrocarbons derived from oil and natural gas occupy a central position in the chemical industry, forming the basis of many synthetic materials and essential chemical products. Among aliphatic hydrocarbons, ethylene and propylene are particularly important, as they are the basic raw materials for the production of polymers such as polyethylene and polypropylene. These polymers are ubiquitous in our everyday lives: polyethylene, for example, is used in packaging films, plastic bags, bottles and food containers, while polypropylene is preferred for textiles, fibers, automotive components and medical devices due to its mechanical and chemical resistance (4).

Aromatic hydrocarbons play an important role in synthetic organic chemistry. Benzene, toluene and xylene, obtained mainly by catalytic reforming, are precursors for a wide range of products. Benzene is used to manufacture solvents, resins and pharmaceuticals. It is also the raw material for the synthesis of numerous compounds such as phenol, used in the manufacture of phenolic resins, and aniline, used in the production of dyes and synthetic rubbers. Toluene, for its part, is used as an industrial solvent, but also as a base for the manufacture of trinitrotoluene (TNT) and diisocyanates, necessary for the production of polyurethanes. Xylene, used in adhesives and inks, is also the raw material for dimethyl terephthalate, a precursor of polyester fibers.

In addition to synthetic materials and solvents, hydrocarbons are essential for the production of detergents and cleaning products. Alkylbenzenes, derived from benzene, are transformed into linear alkylbenzene sulfonates (LABS), which are the surfactants in many household and industrial detergents. In the pharmaceutical industry, aromatic hydrocarbons serve as precursors for active molecules and excipients, enabling the development of drugs and health products.

In short, hydrocarbons form the backbone of the modern chemical industry. They are not only raw materials for plastic products, but also essential compounds in the production of solvents, detergents, medicines, textiles and many other industrial goods. Their versatility and availability make them indispensable mainstays for a wide range of industrial sectors, from the automotive and food industries to healthcare, construction and consumer goods.

II.1 Description of coal and its properties

Hard coal, a type of bituminous coal, is rich in carbon and contains various complex aromatic compounds. It is formed by the decomposition of organic matter under high pressure and temperature, and is distinguished by its high energy density. Hard coal is a valuable source of carbon, as well as many other compounds during distillation. Its chemical properties make it a versatile raw material for producing gas, aromatic chemicals and coke, an essential material for the metallurgical industries.

II.2 Coal distillation: town gas, tar, coke

High-temperature coal distillation produces several useful by-products:

II.2.1. Town gas

Mainly made up of methane, hydrogen and carbon monoxide, it has been used for decades as a source of district heating and street lighting.

II.2.2. Tar

Coal tar is a rich source of aromatic compounds such as benzene, toluene and xylenes. These compounds are the building blocks of aromatic organic chemistry, and are used to manufacture dyes, solvents and resins.

II.2.3. Coke

Coke is a carbon-rich solid residue used as a fuel in metallurgical furnaces and as a reducing agent in iron and steel production. It is also essential in other metal treatment processes due to its high carbon purity (6).

II.3. Applications of coal distillation by-products

The by-products of coal distillation have a wide range of applications. Benzene and toluene, for example, are used as raw materials in the manufacture of plastics, paints, detergents and medicines. Coke is used in the steel industry to produce steel and for various foundry applications, while town gas, although less widely used today, has played a key role in lighting and heating for decades. Carbochemicals, which use these by-products, form the basis of a wide range of industrial products and remain an important alternative to petroleum hydrocarbons.

III. Biomass and Green Chemistry

III.1 Biomass sources: cellulose, lignin, vegetable oils

Biomass is a renewable raw material derived from plant and animal sources. It includes compounds such as cellulose, lignin and vegetable oils. **Cellulose** is a natural polymer extracted from plants and forms the basis of bioplastics and biofuels. **Lignin**, meanwhile, is a component of wood used to manufacture adhesives and resins. Finally, **vegetable oils** from plants such as soy and rapeseed can be transformed into biodiesel and biosourced plastics (7).

III.2 Conversion techniques: fermentation, pyrolysis

Biomass is transformed into chemical products by various processes:

III.2.1. Fermentation

Uses micro-organisms to convert sugars into ethanol, lactic acid or biogas. These products are essential for the biofuels and biodegradable plastics industries.

III.2.2. Pyrolysis

Thermal decomposition of biomass produces bio-oils, biochar and gases. These products can be used as substitutes for hydrocarbons in energy production and in certain industrial applications.

III.3. Recent advances in green and sustainable chemistry

Green chemistry, which aims to reduce the environmental impact of chemical processes, has made significant progress with the development of **green solvents**, **enzymatic catalysts** and **biodegradable** materials. Advances in bioengineering and synthetic biology are enabling the production of bio-sourced substances to replace plastics and traditional chemicals, supporting a transition to a more sustainable and environmentally-friendly chemical industry.

IV. Comparison of raw material sources

IV.1 Advantages and disadvantages of each source

IV.1.1. Oil and Natural Gas

Oil and natural gas are the world's most widely used energy sources, due to their abundance and high energy yield. These fossil resources are essential not only as fuels, but also as raw materials in the production of chemicals, plastics and many other synthetic materials. Petroleum, for example, is a basic raw material for products such as gasoline, diesel, lubricants and plastics, while natural gas is an important source for the production of ammonia, methanol and hydrogen.

However, the **non-renewable** nature of oil and natural gas raises sustainability issues. Available reserves are limited and concentrated in certain regions of the world, creating geopolitical issues linked to energy security. Growing demand for these resources, particularly in developing countries, could accelerate their depletion, leading to price instability and tensions between importing and exporting countries.

What's more, these resources have a major environmental impact due to the **greenhouse gas emissions** they generate. When burned, oil and natural gas release carbon dioxide (CO_2), a greenhouse gas contributing to global warming. The processes of extracting, refining, and transporting oil can also lead to leaks and spills, causing damage to marine and terrestrial ecosystems. Dependence on these fossil fuels is a challenge for climate policies, prompting governments to seek renewable alternatives and encourage innovation in clean energy.

VI.1.2. Hard coal

Hard coal, or bituminous coal, was one of the first fossil fuel sources used to fuel industrialization in the 19th century, playing a central role in the development of heavy industry. Rich in carbon, hard coal is used today in specific applications, notably in the **steel** and carbochemical industries. In steel production, for example, coke derived from coal is indispensable as a reducing agent, enabling the transformation of iron ore into metallic iron. Hard coal is also used in electricity generation, although its use is declining in this field due to its environmental impact.

Hard coal's main drawback is its **high carbon footprint**. When burned, it emits large quantities of CO_2, as well as other air pollutants such as sulfur oxides (SOx) and nitrogen oxides (NOx), responsible for soil acidification and acid rain. It also emits fine particles, which have a detrimental effect on air quality and public health, causing respiratory and cardiovascular disease.

What's more, coal mining is an activity with a high ecological impact: open-cast mines degrade landscapes, destroy natural habitats and contaminate groundwater. Acidic water discharges from mines also pollute watercourses, affecting aquatic biodiversity. Despite its usefulness in certain industries, hard coal is considered one of the most polluting energy sources, prompting many countries to restrict or abandon its use for environmental reasons.

VI.1.3. Biomass

Biomass is a renewable alternative to fossil fuels, obtained from **organic matter** such as wood, agricultural waste, algae, and vegetable oils. Unlike fossil resources, biomass is considered a **carbon-neutral** energy source: the CO_2 emitted during its combustion is offset by the CO_2 captured by plants as they grow. This makes biomass a more sustainable option for the production of fuels (such as biodiesel and bioethanol) and biobased chemicals.

However, harnessing biomass for industrial applications is not without its challenges. **Processes for converting** biomass into biofuels or chemicals often require costly, energy-intensive technologies, such as fermentation, pyrolysis or gasification. These processes also require a significant input of raw materials, and large-scale biomass production can lead to **competition for agricultural land**. This could affect the availability of land for food farming, particularly in regions where resources are limited.

In terms of efficiency, biomass is often less efficient than fossil hydrocarbons due to the low energy density of the source materials. For example, converting algae into liquid fuel requires a large amount of raw material to obtain a relatively small volume of fuel. This limits the applicability of biomass in sectors with high energy requirements.

Despite these limitations, biomass continues to play a growing role in the energy transition, thanks to advances in green chemistry and biotechnology. Innovations in the development of bio-reactors, modified micro-organisms and more efficient conversion technologies could improve the profitability and yield of biomass, making this resource a more viable and sustainable alternative for the future.

IV.2 Environmental impact and sustainability issues

The exploitation of hydrocarbons and coal contributes to CO_2 emissions and global warming. Biomass, with a more carbon-neutral balance sheet, reduces this impact, but must be exploited sustainably to avoid excessive pressure on natural resources.

IV.3. Future outlook for each raw material source

The future outlook for organic chemical feedstocks is shaped by environmental challenges, technological innovations and increasingly stringent international policies. Each of the raw material sources - fossil hydrocarbons, coal and biomass - has a role to play in this transition, but the trajectory of each resource varies according to the economic, environmental and technological priorities of countries and companies.

IV.3.1. Fossil fuels: a gradual but inevitable transition to more sustainable alternatives

Fossil hydrocarbons, such as oil and natural gas, should continue to play a role in the coming decades, but their importance is gradually diminishing. Despite advances in renewable energies, fossil hydrocarbons remain indispensable for many basic products, including plastics, aviation fuels and basic chemicals. Nevertheless, several trends indicate that their role will change:

- **Emissions reduction technologies** : Carbon capture and storage (CCS) and carbon capture, utilization and storage (CUSC) technologies are being developed to reduce emissions associated with fossil hydrocarbons. These technologies could enable cleaner use of fossil resources, prolonging their use in specific applications.
- **Optimizing resources**: The petrochemical industry is evolving to maximize chemical yields and reduce waste, with processes such as catalytic reforming and steam cracking more efficient in terms of energy consumption and reduction of harmful by-products.
- **Divestment and regulation**: More and more financial institutions and governments are introducing restrictions on the financing of fossil fuel-based projects, encouraging divestment in these sectors. Stricter regulations, such as those of the European Union to reduce greenhouse gas emissions, are pushing companies to turn to renewable energies and raw materials.

Although fossil hydrocarbons remain indispensable in certain industrial sectors, the energy transition and technological advances are limiting their role in the medium term.

Industries and governments are investing in alternatives, and environmental policies continue to encourage the transition to more sustainable sources.

IV.3.2. Hard coal: A declining but strategic role in certain industrial sectors

Coal, particularly in the form of coke for the iron and steel industry, plays a vital role in certain industrial sectors, but is one of the most polluting resources. The decline of hard coal is particularly marked in electricity generation, where it has been replaced by renewable energies and natural gas in many countries. However, for certain applications, notably metallurgy, efforts are being made to reduce its environmental impact:

- **Decarbonization of the steel industry**: The steel industry, a major consumer of coke, is looking for alternative methods, such as the use of hydrogen to reduce iron ore. Initiatives such as Sweden's HYBRIT project aim to replace coke with hydrogen, creating carbon-free steel.
- **Emissions capture technologies**: For industries that still rely on coke, CO_2 capture technologies and fine particle filters are being introduced to reduce the environmental impact of emissions. These technologies enable plants to continue operating while complying with stricter environmental standards.
- **Recycling and circular economies**: As part of the circular economy, recycled steel reduces dependence on coke and, consequently, on hard coal. Investment in steel recycling reduces the need to extract and process new raw materials.

In short, although its role is diminishing, coal retains a strategic place in certain heavy industries. However, pressure to decarbonize industrial processes could gradually reduce its demand, encouraging innovations that favor more environmentally-friendly materials and processes.

IV.3.3. Biomass and green chemistry: Towards a significant increase in production and use

Biomass represents the future of the organic chemical industry as part of the transition to a low-carbon economy. Biobased materials, derived from renewable resources, enjoy the advantage of not increasing CO_2 emissions, provided their production is managed sustainably. Prospects for biomass and green chemistry are expanding rapidly thanks to advances in several areas:

- **Improved conversion processes**: Fermentation, pyrolysis and gasification technologies are becoming increasingly efficient, enabling biomass to be converted into a wide range of chemical products. Enzymatic catalysts and biotechnologies are making conversion processes less energy-intensive and more selective, thereby increasing the economic viability of biomass.

- **Development of biobased materials**: Bioplastics and biodegradable materials are gaining ground, particularly in the packaging, automotive and textile industries. Biopolymers such as PLA (polylactic acid) and PHA (polyhydroxyalkanoate) are increasingly competitive alternatives to traditional plastics.
- **Production of advanced biofuels**: Research is focusing on second-generation biofuels, using non-food sources such as agricultural residues and wood waste. These biofuels limit competition for agricultural land and reduce environmental impact compared with first-generation biofuels.
- **Policies and incentives**: Many governments actively support green chemistry and bio-based materials through subsidies, tax credits and sustainability targets. The European Union, for example, has put in place policies favoring renewable materials as part of the Green Deal, aiming to neutralize its carbon emissions by 2050.

References

1. Yergin, D. (1991). *The Prize: The Epic Quest for Oil, Money, and Power*. Free Press.
2. Speight, J. G. (2007). *The Chemistry and Technology of Petroleum*. CRC Press.
3. BP (2023). *Statistical Review of World Energy*.
4. Tissot, B. P., & Welte, D. H. (1984). *Petroleum Formation and Occurrence*. Springer.
5. Matar, S., & Hatch, L. F. (2000). *Chemicals: Essential Chemicals for Industrial Processes*. Gulf Professional Publishing.
6. Jones, D. S. J., & Pujadó, P. R. (2006). *Handbook of Petroleum Processing*. Springer.
7. Gary, J. H., Handwerk, G. E., & Kaiser, M. J. (2007). *Petroleum Refining: Technology and Economics*. CRC Press.
8. Meyers, R. A. (Ed.). (2004). *Handbook of Petrochemicals Production Processes*. McGraw-Hill.
9. Speight, J. G. (2014). *The Refinery of the Future*. Elsevier.
10. Kirk-Othmer (2004). *Kirk-Othmer Encyclopedia of Chemical Technology* (5th ed.). Wiley-Interscience.

Chapter III: Transformation Processes in Chemistry

The third chapter describes the main transformation processes, such as hydration, oxidation and polymerization of alkenes, as well as treatment processes for aromatic hydrocarbons. It also covers key reactions such as reforming, chlorination and oxocompound synthesis. This chapter provides an overview of modern methods for producing high value-added chemical compounds, essential in many industrial sectors.

I. Ethylene and other alkenes

I.1 Hydration: Production of alcohols

The hydration of alkenes, in particular ethylene, is an imprtant process for obtaining alcohols, of which ethanol is the most common. The process involves adding a molecule of water to the alkene in vapor form in the presence of an acid catalyst, often dilute sulfuric acid or a zeolite-type catalyst (1). The ethanol produced is essential not only as a solvent and alternative fuel, but also as a raw material for other chemical transformations, such as the production of ethylene glycol or acetaldehyde (2). This process can also be applied to other alkenes to produce various alcohols, which are synthetic intermediates in the manufacture of cosmetics, personal care products and many industrial products.

I.2 Oxidation: Production of ethylene oxide and ethylene glycol

The oxidation of ethylene to ethylene oxide is a key catalytic process in the chemical industry, usually carried out in the presence of silver-based catalysts. Ethylene oxide is a highly reactive molecule, which can be converted into ethylene glycol, a liquid used for its antifreeze properties in automotive and industrial cooling systems. Ethylene glycol is also an indispensable intermediary in the manufacture of polyethylene terephthalate (PET) (4), an omnipresent polymer in textiles (polyester fibers) and food packaging (plastic bottles). The conversion of ethylene oxide into ethylene glycol represents a strategic sector in organic chemistry, as it enables the manufacture of essential materials from simple products such as ethylene.

I.3 Polymerization: Production of polyethylene and other polymers

Ethylene polymerization, carried out by radical or catalytic polymerization, produces various types of polyethylene, each with specific properties suited to distinct uses. For example, low-density polyethylene (LDPE) is used for plastic bags and packaging films because of its flexibility (5), while high-density polyethylene (HDPE) is preferred for pipes, tanks and containers, thanks to its rigidity and mechanical strength. The production of polypropylene (PP) and other alkene polymers, obtained by copolymerization or using different catalysts (such as Ziegler-Natta catalysts), extends the use of alkenes to create materials with diversified properties, indispensable in the construction, automotive and electronics sectors.

Table 3: Main ethylene and derivatives

Matière de base	Intermédiaires	Produits finis. Utilisations
ÉTHYLÈNE 3 400 000 t/an(*)	→ Oxyde d'éthylène → Glycol (Ethane-1,2-diol)	*Textiles polyesters, antigels, explosifs.*
	→ Ethers de glycol	*Solvants, peintures, vernis, liquides hydrauliques.*
	→ Ethanolamines	*Agents tensioactifs, cosmétiques.*
	→ Polyéthers	*Pharmacie, cosmétiques, détergents.*
	→ Éthanol	*Solvants, synthèse.*
	→ Éthanal → Chloral	*Colorants, insecticides, (DDT).*
	→ Acide acétique	*Rayonne, insecticides, rhodoïd.*
	→ Aldol, butadiène	*Élastomères.*
	→ Dérivés chlorés	*Solvants dégraissants.*
	→ Chlorure de vinyle	*Chlorure de polyvinyle, objets moulés, feuilles.*
	→ Acétate de vinyle	*Acétate de polyvinyle, peinture, adhésifs.*
	→ (Polymérisation)	*Polyéthylènes, objets moulés, feuilles, films.*
	→ Éthylbenzène → Styrène	*(voir dérivés du benzène).*
PROPYLÈNE ou PROPÈNE 2 400 000 t/an(*)	→ Oxyde de propylène → Propylèneglycol (Propane-1,2-diol)	*Solvants, antigels, fluides hydrauliques, parfumerie, détergents, polyuréthanes.*
	→ Alcools « Oxo »	*Solvants synthèse.*
	→ Propan-2-ol	*Synthèse, acétone, solvant.*
	→ Acrylonitrile	*Fibres textiles, résines élastomères, peintures.*
	→ Tétrachlorométhane	*Dérivés fluorés et chlorofluorés, solvant.*
	→ Chlorure d'allyle → Glycérol	*Résines glycérophtaliques, peintures, explosifs, pharmacie.*
	→ (Polymérisation)	*Polypropylène, résines thermoplasiques, plastifiants, détergents.*
	→ Cumène → Phénol + Acétone	*(voir dérivés du benzène).*

- ***Preparation of acetaldehyde***: previously made from acetylene (by hydration), acetaldehyde is now produced more economically from ethylene, using the following reactions:

$$C_2H_4 + PdCl_2 + H_2O \longrightarrow CH_3 - CH = O + Pd + 2\ HCl$$

$$Pd + 2\ CuCl_2 \longrightarrow PdCl_2 + 2\ CuCl$$

$$2\ CuCl + 1/2\ O_2 + 2\ HCl \longrightarrow 2\ CuCl_2 + H_2O$$

lont le bilan se réduit à :

$$C_2H_4 + 1/2\ O_2 \longrightarrow CH_3 - CH = O \qquad (\Delta H = -243\ kJ/mol)$$

The process is exothermic overall, whereas acetylene production, the first stage in the old process, is highly endothermic and therefore energy-intensive.

- ***Preparation of acrylonitrile:*** CH_2=CH-C≡ N. This is a propene-specific reaction, in the presence of ammonia and oxygen; acrylonitrile is polymerizable (synthetic textiles).

II. Benzene and aromatic hydrocarbons

II.1 Alkylation of benzene: ethylbenzene, cumene

Benzene alkylation is a process that introduces an alkyl group into the aromatic structure of benzene, enabling the production of derivatives such as ethylbenzene and cumene, which are crucial intermediates for the chemical industry. Ethylbenzene is converted into styrene, an essential monomer for polystyrene, a lightweight, insulating material used in packaging and building materials. Cumene is converted to phenol and acetone via the process of peroxidation. Phenol is a raw material for phenolic resins and polycarbonates, while acetone is used as a solvent in varnishes, inks and adhesives.

- ***Benzene is alkylated*** by condensation with alkenes (Friedel-Crafts type reaction) using :
    - *Ethylene* to give ethylbenzene, dehydrogenation of which yields polymerizable styrene C6H5 -CH=CH2 (8).
    - *Propene*, to obtain cumene (or isopropylbenzene), an intermediate in the simultaneous preparation of two very important derivatives: *phenol* and *acetone* (6).

Note

Phenol is used in a wide variety of manufacturing processes: high polymers (phenol-formaldehyde resins), dyes (by reaction with diazonium salts), insecticides, fungicides, explosives (picric acid), nylon, medicines, etc.

Annual world consumption is around 4 million tonnes, and a single cumene process plant can produce 200,000 tonnes a year, or over 500 tonnes a day.

Acetone is a solvent and an intermediate in various syntheses.

-
    - *Medium-molecular alkenes* for subsequent sulfonation of the resulting alkylbenzenes (e.g. dodecylbenzene C6H5-C12H25) to produce synthetic detergents.

- **Benzene *chlorination*** produces either addition derivatives such as hexachlorocyclohexane, used as an insecticide, or substitution derivatives such as monochlorobenzene, which is used with chloral $CCl_{(3)}$-CH=O in the manufacture of D.D.T. (*Dichloro-Diphenyl-Trichloroethane*):

$$2\,C_6H_5Cl + CCl_3-CH{=}O \rightarrow Cl-C_6H_4-\underset{\displaystyle CCl_3}{\underset{|}{CH}}-C_6H_4-Cl$$

Nitration of benzene or toluene produces mononitro derivatives, the reduction of which yields the corresponding amines, raw materials in dye chemistry; trinitro derivatives are also explosives, such as trini trotoluene (T.N.T.) (6).

Oxidation of benzene hydrocarbons yields various acids, notably diacids used in polycondensation reactions: maleic acid from benzene, ortho-phthalic acid from *ortho-xylene* or naphthalene, terephthalic acid from *para-xylene*.

- **Benzene *hydrogenation*** is a route to cyclohexane, whose oxidation to a mixture of cyclohexanol and cyclohexanone leads to either adipic acid or caprolactam, intermediates in the manufacture of nylon 6/6 and nylon 6.

II.2 Aromatics applications: phenol, acetone, styrene

Because of their stable, reactive chemical structure, aromatic compounds have a wide range of applications. Phenol is a precursor of synthetic resins such as phenolic resins and epoxy resins, used for adhesives and thermal insulation. Acetone, beyond its role as a solvent, is also used in the manufacture of methyl methacrylate (MMA) and acrylic plastics, while styrene is the basis of polystyrene and styrene-butadiene rubber (SBR), impregnated for tires and foam products (9).

II.3 Importance of nitro and chlorine derivatives in industrial chemistry

Nitro and chlorine derivatives of aromatics are indispensable chemical intermediates. For example, nitrobenzene is a precursor of aniline, used to produce polyurethanes and dyes. Chlorinated compounds, such as chlorobenzene, are used as solvents and intermediates in the manufacture of agrochemicals, pharmaceuticals and crop protection products. These derivatives play a critical role in the synthesis of complex chemicals, often being more reactive and scalable for specific applications in the chemical industry (10).

III. Main chemical reactions

III.1 Chlorination: halogenated derivatives and solvents

Chlorination of hydrocarbons produces halogenated derivatives such as carbon tetrachloride and methylene chloride, which are powerful solvents for the paint, coatings and cleaning products industries. These halogenated derivatives are also used as degreasing agents in the mechanical and electronics industries. Vinyl chloride, produced by the chlorination of ethylene, is the basic monomer for PVC, a polymer used in pipes, electrical cables and floor coverings.

III.2 Reforming: improvements in octane number

Catalytic reforming improves fuel quality by increasing octane number, which in turn improves fuel resistance to detonation. This process, carried out in the presence of platinum- or rhodium-based catalysts, converts naphthas into aromatic hydrocarbons such as toluene and xylene. These aromatics are used to improve the performance of fuels while producing raw materials for the petrochemical industry.

III.3 Oxo-synthesis: Production of aldehydes and alcohols

Oxo-synthesis, or hydroformylation, adds a formyl group to alkenes to produce aldehydes, such as butyraldehyde, which can then be hydrogenated to produce alcohols. These alcohols are used in the manufacture of polymer plasticizers, fuel additives and as solvents in industrial products. This reaction is generally catalyzed by transition metal complexes, such as cobalt or rhodium, and represents a key step in the chemistry of aldehydes and alcohols.

IV. Polymerization and polycondensation

IV.1 Addition polymerization: Polyethylene, PVC, Polystyrene

Addition polymerization involves joining identical monomers without releasing by-products. Polyethylene, produced by the polymerization of ethylene, is used in plastic films, bottles and food containers because of its chemical resistance and light weight. PVC, produced from vinyl chloride, is prized for its flame-retardant properties and weather resistance, making it ideal for construction applications. Polystyrene, meanwhile, is used in packaging, insulation and everyday consumer products thanks to its transparency and rigidity.

IV.2 Polycondensation: Polyesters, Polyamides, Phenoplasts, Polyurethanes

Polycondensation involves the reaction between two types of monomer, releasing by-products such as water. Polyesters, such as PET, are used in textiles and packaging, while polyamides, such as nylon, are renowned for their abrasion resistance, making them ideal for technical textiles and mechanical components. Phenoplasts, thermosetting resins, are commonly used in electrical applications and industrial adhesives.

Polyurethanes, produced by the reaction between polyols and isocyanates, are versatile and used in flexible foams, paints and resistant coatings.

IV.3. Polymer applications and their specific properties: Thermoplastics, Thermosets, Elastomers

- **Thermoplastics**: These are polymers that can be melted and reshaped several times, such as polyethylene and polypropylene, widely used in packaging, automotive and consumer products.
- **Thermosets**: These polymers harden irreversibly with heat, like epoxy resins and phenolic resins, used for their chemical and thermal resistance in adhesives and building materials.
- **Elastomers**: These flexible polymers, such as natural rubber and polybutadiene, are prized for their elasticity and resistance to stress. They are used in tires, sealants and industrial gaskets.

The polymers produced by polymerization and polycondensation are varied and meet specific needs in the industrial, transport, electronics and consumer goods sectors, making this sector an essential pillar of organic chemistry.

IV.4. High polymers

High polymers" *are macromolecules*, i.e. molecules with a very high molecular weight (up to a million), and therefore a very large number of atoms.

There are macromolecules in nature, such as cellulose, which is a constituent of plants and wood, proteins (e.g. keratin, which makes up hair and wool), and latex, which is formed by the coagulation of rubber tree sap. Man first sought to chemically modify these natural materials, to adapt them to particular uses: the vulcanization of natural rubber (discovered by Goodyear in 1839), the transformation of cellulose into celluloid (1870) or "artificial silk" or "rayon" fibers (1889) are all part of this approach. Then came the synthesis of new macromolecular materials, known as "high polymers": bakelite (1909), polychloroprene rubber (1931), polyamide fibers (nylon, 1935).

Today, these high polymers have diversified enormously and are used (under the name of "plastics") in countless applications. Their interest lies in the diversity of their properties, and the possibility of preparing materials adapted to each particular use. They can be flexible or rigid, transparent or opaque; they can be used to manufacture large, complicated parts directly and economically by molding; they are usually insulating, but we know how to prepare conductive polymers; their mechanical strength can be improved by combining them with fibers (glass, carbon, etc.) in composite materials.

France is Europe's second-largest producer of high polymers. In the space of a few decades, production has increased by a factor of around 50, to around 6.5 million tonnes a year. Annual consumption in France continues to rise, at around 80 kg per capita.

Synthetic macromolecular compounds are always formed by the *repetition of a simple motif* a very large number of times. The term "*polymer*" is used to describe this type of structure, and the adjective "*high*" refers to their very high molecular weight.

Some of the resulting products are liquids, but most are solids. They are They are "*thermoplastic*" if they can be softened by heat and regain their normal hardness on cooling (making it possible to manufacture objects by molding, after the chemical elaboration of the polymer); they are "*thermosetting*" if the reaction that produces them leads to a solid mass that can no longer be softened by heat (need to carry out the reaction in the mold). Finally, some polymers have the elastic properties of rubber; these are known as "*elastomers*".

These molecules can be obtained by "polymerization" or "polycondensation".

IV.1 Polymerization

Polymerization proper, or "addition polymerization", consists in joining together the molecules of a simple compound, called the "monomer", to give, without any elimination, a compound with a higher molecular weight, an integer multiple of that of the monomer. The number of monomer molecules bonded together in this way can be very large (several tens of thousands), and the molecular weight of the polymer can exceed one million; but polymers with molecular weights limited to around twenty-five thousand are also known and in use.

The monomers are always unsaturated compounds, and polymerization has already been mentioned among the possible reactions of alkenes and conjugated dienes. The mechanisms involved have also already been mentioned, so only the two corresponding diagrams will be recalled here:

Table 5: Main addition polymers

Monomer	**Polymer**	**Applications**
Ethylene $H_2C{=}CH_2$	Low-density polyethylene	*Films, foils (packaging, agriculture), household objects, cables.*

	High density	*Molded objects, bottles, hollow bodies.*
Propene $CH_{(3)}$-CH=CH_2	Polypropylenes	*Moulded articles (automotive), furniture, sanitaryware, cables).*
Vinyl chloride CH_2=CHCl	Polyvinyl chloride	*Rigid: pipes, electrical sheaths, bottles, hollow bodies* *Flexible (plasticized): films and sheets, floor coverings, cables, shoes, toys.*
Styrene C_6H_5 -CH=CH_2	Polystyrene	*Containers, furnishings, toys.* *Expanded (98% air): impact-resistant packaging, heat and sound insulation.*
Acrylonitrile CH_2=CH-CʃN	Polyacrylonitrile	*Acrylic" textile fibers (Orlon)*
Methyl metharcrylate CH_2 =C(CH_3)CO_2CH_3	Polymeth/ acrylic	*Organic glass (Plexiglas, Altuglas).*
Tetrafluoroethylene F_2C=CF_2	Polytetrafluoroethylene (Teflon)	*Heat-resistant coatings, mechanical parts, high chemical resistance materials*
Butadiene H_2C=CH-CH=CH_2 Isoprene H_2C=C(CH_3)-H=CH_2 Chloroprene H_2C=CCl-CH=CH_2	Various elastomers (often copoly- mothers, e.g. butadiene-styrene)	*Tires, tubes, hoses; various rubber applications.*

VI.1.1. Alkenes and vinyl compounds

$$n\,A-CH=CH_2 \rightarrow \left(\underset{A}{\underset{|}{CH}}-CH_2\right)_n$$

Examples

Ethylene (A=H), propene or "propylene" (A=CH_3), vinyl chloride (A=Cl), styrene (A=Ph), acrylonitrile (A=CN), etc.

VI.1.2. Conjugated diene

$$n\,H_2C=\underset{A}{\underset{|}{C}}-CH=CH_2 \rightarrow \left(CH_2-\underset{A}{\underset{|}{C}}=CH-CH_2\right)_n$$

Examples

Buta-1,3-diene (A=H), chloroprene (A=Cl).

The polymers obtained in this way have a linear main chain. They are either thermoplastics (ethylene and vinyl compounds) or elastomers (dienes).

From a stereochemical point of view, these polymers can have several geometric arrangements. There are in fact three, which differ in the arrangement of the A groups relative to the main carbon chain, assumed to be "elongated" in a zigzag conformation:

- ***Isotactic polymers***, where the A groups are all "on the same side":

- ***Syndiotactic polymers***, where the A groups are alternately on one side and on the other:

- ***Atactic polymers***, where the A groups are randomly distributed between the two sides:

These geometric differences lead to significant differences in properties (melting or softening points, tensile strength, etc.). Special catalysts can be used to prepare one or other of these forms. Frequently, to modify the polymer's properties and adapt them to specific uses, a blend of different polymers is "copolymerized" (e.g. a mixture of butadiene and styrene, for certain synthetic rubbers).

Finally, a "polyaddition" reaction characteristic of epoxies, such as ethylene oxide, is generally associated with polymerization reactions. *"polyaddition"* reaction characteristic of epoxides such as ethylene oxide. In the presence of an alcohol or phenol, and under the effect of acid or base catalysis, the following reaction can occur:

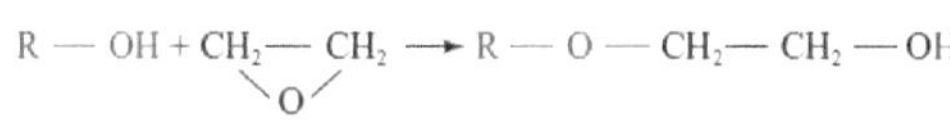

The resulting product is both an ether and an alcohol, and the latter function lends itself to a further identical reaction with a second molecule of ethylene oxide to give R-O-$CH_{(2)}$)-$CH_{(2)}$)-O-$CH_{(2)}$)-CH_2OH.

The reaction can then continue to lead to "polyethers" $R(OCH_2CH_2)_nOH$, the initial alcohol having served only as an initiator.

VI.2 Polycondensation

This is a genuine chemical reaction between molecules carrying different functional groups (e.g. esterification between alcohols and acids), but as the reactants carry two (or three) functional groups, a step-by-step reaction gradually builds up very long chains.

The "antagonistic" functions (represented below by X and Y) can be carried by different molecules:

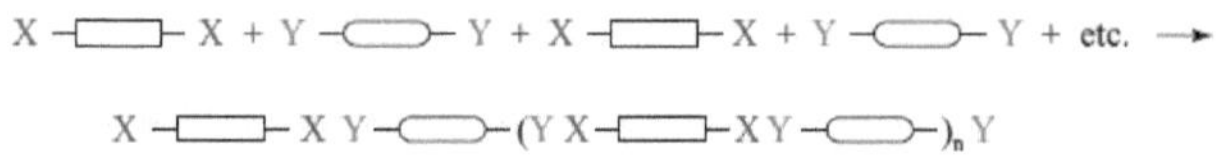

or be present at both ends of the same molecule:

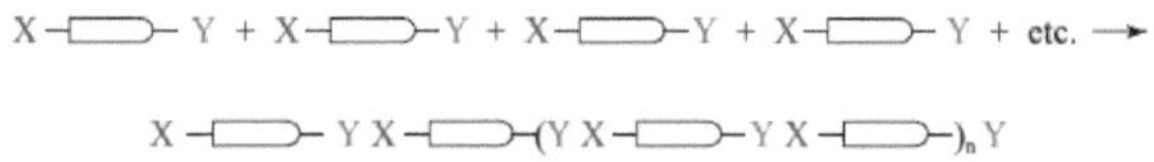

The linear products thus formed are thermoplastic, but if one of the reactants has three functions (or three reaction sites), trans- versal chains can be established, forming "bridges" between the linear chains and creating a three-dimensional network, corresponding to a thermosetting product:

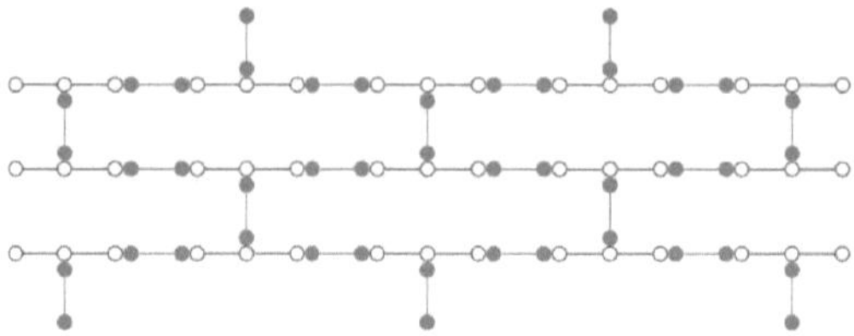

Figure 2: Structure of a three-dimensional resin

The reaction of the trifunctional reagent with its two terminal functions first forms linear chains. These are then joined by "bridges", which are established between them by the central function. The diagram opposite is two-dimensional, but these bridges can also link chains that are not all in the same plane (here, that of the paper), to form a three-dimensional network.

Par simplification, 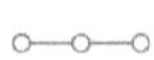 représente ici le composé trifonctionnel 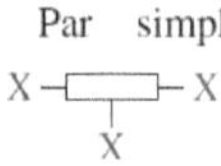 et  représente le composé bifonctionnel 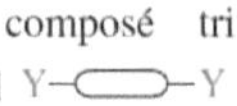

The products formed, which are "polycondensates", are often also referred to as "polymers" (or "condensation polymers").
They are often also called "polymers" (or "condensation polymers"), although these expressions are not correct, since they do not result from a polymerization reaction. Those with a linear structure (exclusively bifunctional reactants) are thermoplastics; they are mainly used in the production of synthetic textile fibers, by melt-spinning. Those with a three-dimensional structure (one of the trifunctional reagents) are thermosetting "resins".

VI.2.1. Polyesters

They result from the reaction between a diacid and a di (or tri) alcohol, so that the monomer units are joined by ester functions:

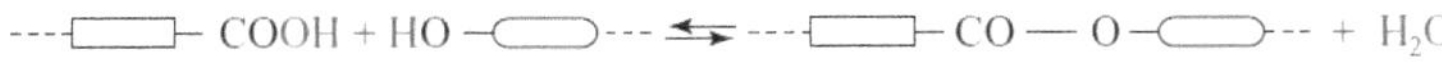

Acide **Alcool** **Ester**

Examples

- Condensation of a *dialcohol* and a *diacid* :

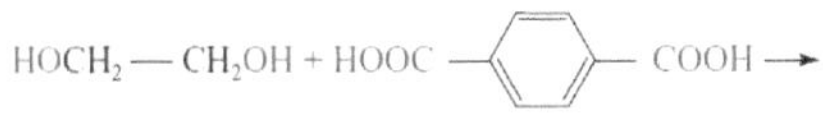

Fibres textiles synthétiques (tergal, dacron, térylène).

Éthane-1,2-diol **Acide téréphtalique**

- Condensation of a dialcohol and a diacid :

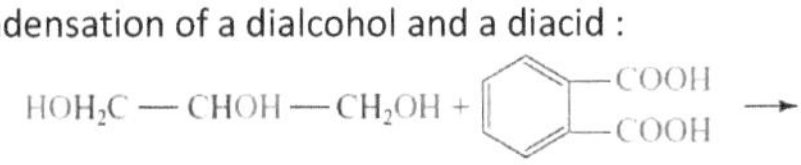

Résines glycérophtaliques (peintures, vernis).

Glycérol **Acide ortophtalique**

The materials used in these two examples are obtained from ethylene, propene, *ortho-* and *paraxylene.*

VI.2.2. Polyamides

The base reaction takes place between acid and amine functions, leading to amide functions:

Acide **Amine** **Amide**

Examples

- Condensation of a diacid on a diamine :

$$HOOC - (CH_2)_4 - COOH + H_2N - (CH_2)_6 - NH_2 \rightarrow$$ Nylon 6/6 (fibres textiles, objets moulés).

Acide adipique **Hexaméthylènediamine**

Kevlar (matériaux légers et de très grande résistance mécanique).

***p*-phénylènediamine** **Chlorure de téréphtalyle**

- Condensation of an amino acid on itself :

Note

6/6 means "formed from two reactants each containing six carbons". 6 means "formed from a single six-carbon reactant".

$H_2N - (CH_2)_5 - COOH$ **Acide 6-aminocaproïque** $\rightarrow$ Nylon 6, Perlon (fibres textiles).

$H_2N - (CH_2)_{10} - COOH$ **Acide 11-aminoundécanoïque** $\rightarrow$ Rilsan, (fibres textiles).

Of the many industrial methods used to obtain the molecules needed for these condensations, only the "chain" using benzene as the base material will be mentioned here.

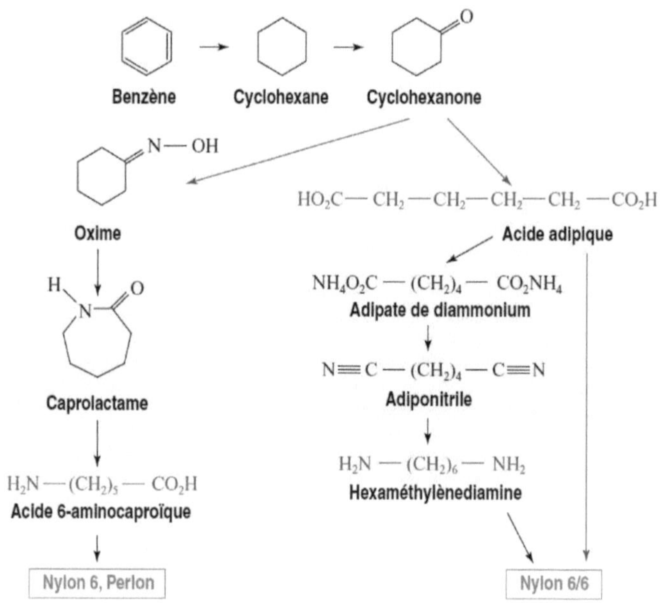

11-aminoundecanoic acid is obtained from ethylenic acid $H_2C{=}CH{-}(CH_2)_8{-}COOH$, which in turn results from the thermal cracking of castor oil.

VI.2.3. Phenoplasts and aminoplasts

The formation of these thermosetting products involves a reaction between formaldehyde $H_2C{=}O$ and a labile hydrogen compound, as follows:

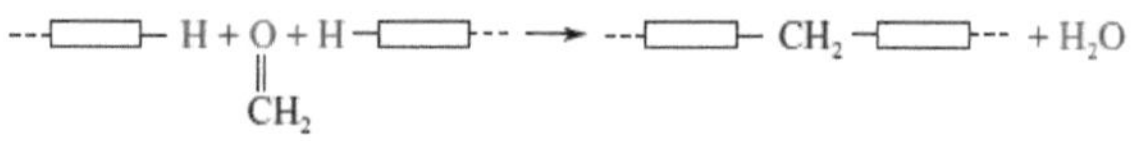

Phenoplast resins result from a reaction of this type with phenol, which then has three reactive sites: the two ortho positions and the para position. Production of these resins totalled around 75,000 tonnes in 2003.

In the synthesis of *aminoplast resins* (176,000 tonnes in 2003), the labile hydrogen compound may be urea $H_2N{-}CO{-}NH_2$, whose four hydrogen atoms are liable to react, forming a three-dimensional network with particularly tight meshes:

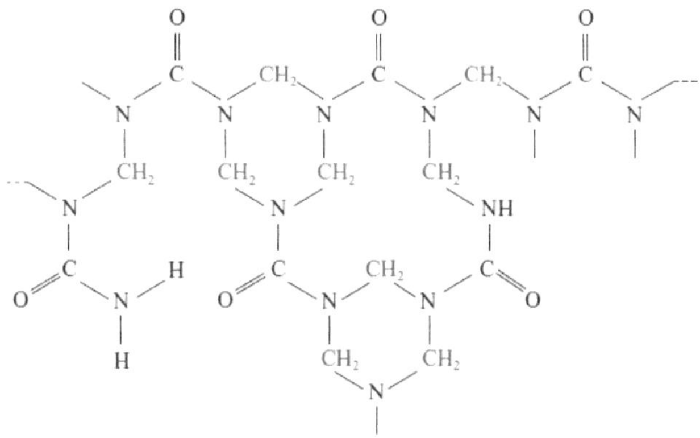

Another nitrogen compound used is melamine, which has six labile hydro-genes ("hardener" glues, "plastic" tableware, etc.).

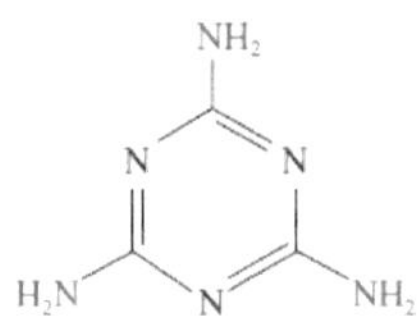

Melamine

VI.2.4. Polyurethanes

Polyurethane synthesis is based on the reaction between an alcohol function and an isocyanate function to form a *urethane*:

$$\underset{\textbf{Isocyanate}}{R-N=C=O} + R'OH \longrightarrow \underset{\textbf{Uréthane}}{R-NH-\overset{\displaystyle O}{\overset{\|}{C}}-OR'}$$

With a dialcohol and a *diisocynate*, polycondensation is possible.

Polyurethanes are thermoplastic solids that can be used to manufacture molded objects (e.g. ski boots). But if polycondensation is carried out in the presence of water, which reacts with the isocyanates to release carbon dioxide gas, the result is a cellular material, either hard or soft and elastic. The latter is known as "plastic foam", widely used as insulation, as well as in seats, mattresses, etc.

Example

Toluene-2,*4-diisocyanate* and ethane-1,2-diol react as follows:

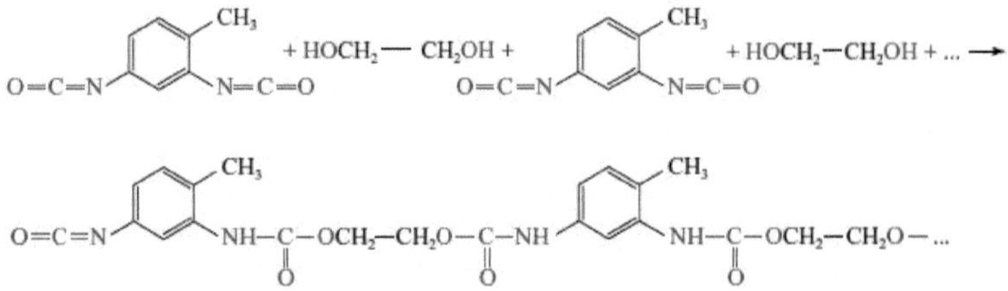

References

1. Satterfield, C. N. (1991). *Heterogeneous Catalysis in Industrial Practice*. McGraw-Hill.
2. Kirk-Othmer (2004). *Kirk-Othmer Encyclopedia of Chemical Technology* (5th ed.). Wiley-Interscience.
3. Meyers, R. A. (Ed.). (2004). *Handbook of Petrochemicals Production Processes*. McGraw-Hill.
4. Jones, D. S. J., & Pujadó, P. R. (2006). *Handbook of Petroleum Processing*. Springer.
5. Speight, J. G. (2007). *The Chemistry and Technology of Petroleum*. CRC Press.
6. McMurry, J. (2015). *Organic Chemistry* (9th ed.). Cengage Learning.
7. Yergin, D. (1991). *The Prize: The Epic Quest for Oil, Money, and Power*. Free Press.
8. Gary, J. H., Handwerk, G. E., & Kaiser, M. J. (2007). *Petroleum Refining: Technology and Economics*. CRC Press.
9. Shreve, R. N., & Brink, J. A. (1984). *Chemical Process Industries*. McGraw-Hill.
10. Matar, S., & Hatch, L. F. (2000). *Chemicals: Essential Chemicals for Industrial Processes*. Gulf Professional Publishing.

Chapter IV:

Applications and Products for the Chemical Industry

This chapter examines products derived from organic chemistry in a variety of fields: parachemistry, pharmaceuticals, consumer goods and building materials. It presents applications for detergents, paints, plastics and pharmaceutical components, highlighting the importance of hydrocarbons for everyday products. The chapter highlights the diversity of applications, from packaging to textiles and from cosmetics to electronic components.

I. Parachemistry products

I.1 Detergents, paints, cosmetics

Detergents are essential cleaning products in households and industry. They contain surfactants - amphiphilic molecules capable of lowering the surface tension of water, effectively dissolving grease and impurities (1). These surfactants are often derived from hydrocarbons by sulfonation or sulfation, giving rise to compounds such as alkylbenzene sulfonates and lauryl sulfates, which enhance cleaning power (2). Additives, such as bleaching agents and enzymes, further enhance detergents' effectiveness (3).

Paints, used to decorate and protect surfaces, are composed of pigments, binders and solvents (4). Organic pigments provide color and opacity, while binders - often acrylic, polyurethane or alkyd resins - ensure adhesion and weather resistance (5). Organic solvents facilitate application by modifying viscosity, then evaporate after drying. This composition enables paints to meet a wide range of requirements, such as durability, UV resistance and corrosion protection.

Finally, cosmetics draw on a wide range of organic compounds to improve the texture, efficacy and stability of skin care and beauty products. Emollients, such as esters and fatty alcohols, soften and moisturize the skin, while preservatives, such as parabens or organic acids, extend product life. Synthetic fragrances provide long-lasting, customizable scents tailored to consumer preferences (6).

I.2 Plant protection products and fertilizers

Phytosanitary products are essential for crop protection. Herbicides, such as glyphosate, inhibit weed growth, while insecticides, such as pyrethroids, target insect pests (7). Fungicides, based on triazoles or strobilurins, protect plants from fungal infections. Organic chemistry makes it possible to design these molecules specifically to minimize their environmental impact, although concerns persist about their effects on biodiversity and human health (8).

Organic fertilizers, such as urea and ammonium nitrate-based compounds, provide the nutrients needed for plant growth, including nitrogen, phosphorus and potassium. These compounds increase crop yields, but the intensive use of fertilizers raises questions about water eutrophication and greenhouse gas emissions. Alternatives, such as controlled-release fertilizers, are being developed to limit these effects (9).

I.3 Explosives, dyes and specific applications

Explosives such as nitroglycerine and trinitrotoluene (TNT) are highly reactive organic compounds used in the construction, mining and defense industries. These compounds release a large amount of energy during their rapid decomposition, enabling rocks or structures to be fragmented efficiently (10).

Organic dyes, on the other hand, are synthetic pigments used to dye textiles, paper and plastics. Azo dyes, for example, are derived from aromatic compounds and produce bright, long-lasting colors suitable for textiles and plastics. Reactive and disperse dyes are also used in industrial and artistic applications for their intensity and stability (1).

Specific applications include niche products such as flame retardants, plasticizers and antioxidants, used to improve the safety and durability of materials in electronics, construction and consumer goods.

II. Pharmaceuticals and Biotechnology

II.1 Importance of chemical intermediates for drugs

Chemical intermediates are essential components in drug synthesis. They serve as the basis for the production of active ingredients, in compliance with purity and specificity standards. For example, amino and carboxylate precursors are used to develop antibiotics, analgesics and antiviral agents. These intermediates are often derived from reactions such as oxidation, reduction and nitration, which transform basic molecules into complex therapeutic compounds (2).

II.2 Innovations in pharmaceutical chemistry and biotechnologies

Innovations in pharmaceutical chemistry are geared towards more sustainable processes, such as enzymatic catalysis and continuous flow chemistry, which enable drugs to be produced with less waste and greater efficiency. Biotechnologies are also playing a growing role in the production of biological drugs, such as monoclonal antibodies and gene therapies, which require sophisticated production processes to maintain the stability and efficacy of biomolecules (3).

II.3 Impact on public health and medical advances

Large-scale drug production has a profound impact on public health, helping to treat infections, chronic diseases and cancers. Advances in pharmaceutical chemistry enable the development of targeted treatments and improved formulations, reducing side effects and increasing patient adherence. In addition, biotechnology research supports the development of vaccines and personalized therapies, opening up new prospects for rare diseases and autoimmune conditions (4).

III. Consumer Products

III.1 Plastics, packaging, textiles

Plastics are ubiquitous in everyday life, used in food packaging, toys, electronic devices and much more. Polymers such as polyethylene (PE), polystyrene (PS) and polyvinyl chloride (PVC) offer properties tailored to the varied needs of consumers. In packaging, plastic films and PET bottles ensure optimum product preservation and reduce transport costs thanks to their light weight (5).

Synthetic textiles, such as polyester and nylon, are made from polymers derived from organic chemistry. These fibers are durable, resistant to wear and tear, and easily modified to create a variety of fabrics to suit the needs of the fashion, furnishing and sports industries.

III.2. Automotive: fuels, plastic components, tires

The automotive industry depends on organic chemistry for many of its key components. Fuels such as petrol and diesel, refined from oil, remain the main sources of energy for combustion-powered vehicles. Plastics, such as polypropylene and ABS, are used in dashboards, bumpers and ventilation systems because of their lightness and impact resistance (6).

Tires, meanwhile, are made from natural and synthetic rubber reinforced with vulcanizing agents. These compounds increase the durability, flexibility and resistance of tires, which have to withstand a wide range of climatic conditions and intensive friction.

III.3 Building materials: resins, paints, insulation materials

Organic chemistry provides essential building materials for durability and safety. Epoxy and polyurethane resins are used in floor coatings, adhesives and sealants for their resistance to water, chemicals and mechanical stress. Insulation materials such as polyurethane foam and expanded polystyrene provide excellent thermal and acoustic insulation, contributing to the energy efficiency of buildings (7).

Protective paints and coatings, which include acrylic resins and anti-corrosion agents, improve the appearance and longevity of materials exposed to the elements. These products help to enhance the durability of infrastructures while meeting growing environmental and safety standards.

References

1. Myers, R. L. (2007). *The 100 Most Important Chemical Compounds: A Reference Guide*. Greenwood Publishing Group.

2. Ullmann, F., Gerhartz, W., Yamamoto, Y. S., Campbell, F. T., & Pfefferkorn, R. (1991). *Ullmann's Encyclopedia of Industrial Chemistry*. Wiley.

3. Kirk-Othmer, (2004). *Kirk-Othmer Encyclopedia of Chemical Technology*, 5th Edition. Wiley-Interscience.

4. Gary, J. H., Handwerk, G. E., & Kaiser, M. J. (2007). *Petroleum Refining: Technology and Economics*. CRC Press.

5. McMurry, J. (2015). *Organic Chemistry*, 9th Edition. Cengage Learning.

6. Speight, J. G. (2007). *The Chemistry and Technology of Petroleum*. CRC Press.

7. Jones, D. S. J., & Pujadó, P. R. (2006). *Handbook of Petroleum Processing*. Springer.

8. Matar, S., & Hatch, L. F. (2000). *Chemicals: Essential Chemicals for Industrial Processes*. Gulf Professional Publishing.

9. Satterfield, C. N. (1991). *Heterogeneous Catalysis in Industrial Practice*. McGraw-Hill.

10. Shreve, R. N., & Brink, J. A. (1984). *Chemical Process Industries*. McGraw-Hill.

Chapter V: Environmental Issues and Challenges for

The fifth chapter looks at the environmental challenges facing the organic chemical industry, such as pollution and waste management. It explores advances towards green chemistry and innovations aimed at reducing the ecological footprint of chemical processes. The chapter also examines international regulations, such as REACH and European standards, and their impact on industrial practices, pushing the industry towards more sustainable practices.

I. Pollution and waste management

I.1 Sources of pollution in the chemical industry

The chemical industry is one of the world's biggest sources of pollution, discharging pollutants into the air, water and soil. Production processes, such as chemical synthesis, distillation, refining and the use of catalysts, generate various polluting by-products (1). For example, the use of organic solvents produces volatile organic compounds (VOCs), which contribute to air pollution and respiratory health problems (2). Sulfur oxides (SOx) and nitrogen oxides (NOx) released by combustion units lead to acid rain, affecting terrestrial and aquatic ecosystems. In addition, heavy metals and toxic chemicals contained in industrial effluents pollute watercourses, disrupting aquatic life (3) and posing risks to human health when these contaminated waters find their way into water supply networks (4).

I.2. Waste treatment and emissions reduction

To counter these effects, the chemical industries are adopting waste treatment and reduction technologies (5). Filtration systems and scrubbers remove solid particles and pollutant gases from the air, while biodegradation and chemical wastewater treatment units treat liquid effluents to minimize their toxicity before discharge (6). In terms of solid waste management, recycling chemical by-products reduces the need for new raw materials and cuts the amount of residue sent to landfill. Solvent recovery processes are also implemented, enabling the reuse of costly solvents while limiting VOC emissions. To take things a step further, some facilities incorporate waste-to-energy units, recovering by-products in the form of heat or electricity (7).

I.3 Air, water and soil pollution

Air pollution from VOC, SOx and NOx emissions causes health problems, including respiratory and cardiovascular disease, and contributes to global warming (8). In water, industrial chemicals disrupt aquatic ecosystems, reduce biodiversity, and contaminate groundwater (9). Heavy metals and persistent organic compounds in soils reduce their fertility and can accumulate in plants, affecting the food chain. Soil remediation techniques, such as phytoremediation (using plants to absorb contaminants) and bioaugmentation (adding microorganisms to break down pollutants), aim to restore contaminated soils. However, these techniques require considerable time and resources, posing a major challenge for chemical industries and governments alike (10).

II. Ecological Transition and Green Chemistry

II.1 Advances in clean and sustainable technologies

The transition to sustainable technologies is one of the priorities of the modern chemical industry (11). Enzymatic catalysis, which uses natural enzymes to carry out chemical transformations, enables complex compounds to be produced with low energy consumption and little waste (12). Continuous-flow synthesis technologies, meanwhile, are becoming increasingly common, enabling chemical compounds to be produced in a controlled manner, using fewer reagents and less energy. Materials chemistry has also developed chemical recycling methods to transform plastics into monomers that can be used in new production cycles, thus reducing dependence on fossil raw materials and the accumulation of plastic waste.

II.2 Use of renewable raw materials

Green chemistry focuses on the use of renewable resources, such as biomass, to produce environmentally-friendly materials and chemicals (13). Vegetable oils are used to produce bioplastics and lubricants, while wood residues such as cellulose and lignin are used to manufacture biodegradable polymers and industrial additives. Research projects are developing second-generation biofuels from non-food sources such as agricultural residues and algae, to avoid competition with food and feed crops. This approach limits greenhouse gas emissions and encourages more sustainable use of natural resources.

II.3 Examples of environmentally-friendly processes

Processes such as biocatalysis and aqueous phase synthesis are examples of environmentally-friendly methods (14). Biocatalysis uses enzymes to catalyze specific reactions, minimizing the use of toxic solvents and waste. Aqueous phase synthesis, which replaces organic solvents with water, reduces pollution risks and waste treatment costs. Microencapsulation, which involves enveloping active substances in biodegradable capsules, enables controlled release and reduces the quantities applied, minimizing environmental impact. These processes demonstrate that it is possible to integrate ecological and cost-effective solutions into industrial production chains.

III. Environmental regulations and standards

III.1 International regulations: REACH, European standards

International regulations, such as REACH (Registration, Evaluation, Authorisation and Restriction of Chemicals) in Europe, require companies to assess and control the risks associated with the chemical substances they produce (15). REACH requires companies to provide safety data for each substance, and bans the use of certain substances deemed hazardous to human health and the environment. In addition, the standards of the European Chemicals Agency (ECHA) restrict the use of certain toxic substances, such as endocrine disrupters, thus encouraging the development of safer alternatives. These regulations aim to protect the environment and consumers by reinforcing the responsibility of chemical producers.

III.2 Impact of environmental policies on industry

Environmental policies directly influence the technological choices and business models of chemical companies (16). Tax incentives and subsidies support projects that incorporate green technologies, while taxes on CO_2 emissions and pollution quotas encourage companies to adopt cleaner production processes. Traceability and transparency obligations, notably via sustainability reports, enable consumers and investors to assess the environmental impacts of companies, favoring those actively committed to reducing their ecological footprint. This trend is leading more and more companies to invest in sustainable solutions and adopt practices in line with the Sustainable Development Goals (SDGs).

III.3 Sustainability requirements for companies

Companies are now required to meet strict sustainability requirements, with targets for greenhouse gas emissions reduction, waste recycling and energy efficiency (17). ISO certifications, such as ISO 14001 for environmental management, attest to companies' efforts to minimize their impact on the environment. Sustainability reports, required by many governments and financial institutions, detail the actions taken to reduce carbon footprints, limit waste and promote the circular economy. These requirements reinforce companies' commitment to sustainable practices, which are becoming essential criteria for business partnerships, market access and global competitiveness. Consumers, increasingly concerned about the environmental impact of the products they buy, favor environmentally-friendly companies, encouraging the whole industry to move towards more responsible models.

References

1. Freeman, H. M. (1988). *Standard Handbook of Hazardous Waste Treatment and Disposal*. McGraw-Hill.
2. Manahan, S. E. (2010). *Environmental Chemistry*. CRC Press.
3. Harrison, R. M., & Hester, R. E. (2002). *Air Pollution and Health*. Royal Society of Chemistry.
4. Alloway, B. J., & Ayres, D. C. (1997). *Chemical Principles of Environmental Pollution*. Blackie Academic & Professional.
5. Baird, C., & Cann, M. (2012). *Environmental chemistry*. W.H. Freeman.
6. Tchobanoglous, G., Theisen, H., & Vigil, S. (1993). *Integrated Solid Waste Management: Engineering Principles and Management Issues*. McGraw-Hill.
7. Clarke, M. J. (1993). *Waste Minimization for Environmental Protection*. Butterworth-Heinemann.
8. Seinfeld, J. H., & Pandis, S. N. (2006). *Atmospheric Chemistry and Physics: From Air Pollution to Climate Change*. John Wiley & Sons.
9. Connell, D. W., & Miller, G. J. (1984). *Chemistry and Ecotoxicology of Pollution*. John Wiley & Sons.
10. Cunningham, S. D., & Berti, W. R. (1993). *Phytoremediation of Contaminated Soils*. Environmental Science & Technology.
11. Sheldon, R. A., Arends, I., & Hanefeld, U. (2007). *Green Chemistry and Catalysis*. Wiley-VCH.
12. Anastas, P. T., & Warner, J. C. (1998). *Green Chemistry: Theory and Practice*. Oxford University Press.
13. Clark, J. H., & Deswarte, F. E. I. (2015). *Introduction to Chemicals from Biomass*. John Wiley & Sons.
14. Hudlicky, T., & Reed, J. W. (2007). *Green Organic Chemistry and its Interdisciplinary Applications*. CRC Press.
15. European Chemicals Agency (ECHA). (2007). *Guidance on Information Requirements and Chemical Safety Assessment*.
16. Porter, M. E., & van der Linde, C. (1995). "Toward a New Conception of the Environment-Competitiveness Relationship". *Journal of Economic Perspectives*, 9(4), 97-118.
17. ISO (2015). *ISO 14001 Environmental Management Systems: Requirements with Guidance for Use*.

Chapter VI :
Perspectives and Innovations in the Chemical Industry

This final chapter focuses on technological innovations and future prospects for the industry. It looks at new materials, nanotechnologies, and the integration of artificial intelligence to optimize processes. The chapter also explores chemical recycling, waste recovery and the principles of the circular economy, illustrating how the chemical industry is evolving to meet tomorrow's economic and environmental challenges.

I. Innovation in Materials and Nanotechnologies

I.1 New materials and smart polymers

New materials developed in organic chemistry offer advanced functionalities, transforming sectors such as healthcare, electronics and transport (1). Among these innovations, **shape-memory polymers** can be programmed to change shape under the effect of external stimuli such as temperature, pH, or light (2). Used in medical devices, they enable the manufacture of stents that expand in blood vessels, or orthoses that adapt to body movements, increasing patient comfort and safety.

Self-repairing materials are another promising area: capable of restoring their mechanical properties after a crack, they extend the life of products, which is particularly advantageous for infrastructure and automotive and aeronautical coatings (3). For example, polymers doped with liquid resin capsules repair themselves when damaged, eliminating the need for frequent maintenance and reducing long-term costs.

Biodegradable polymers, such as PLA (polylactic acid) and PHA (polyhydroxyalkanoate), are increasingly being used to replace conventional plastics in packaging, disposable utensils and even medical devices. Designed to degrade naturally in industrial composting environments, they address growing concerns about plastic pollution and facilitate the transition to a circular economy (4).

I.2 Nanotechnologies and applications in medicine and electronics

Nanotechnologies enable the creation of materials with unique properties at the nanoscale, where specific surface area and reactivity are maximized (5). In medicine, **nanoparticles** offer solutions for targeted treatments (6), where drugs encapsulated in nanoparticles are directed specifically at diseased cells, increasing efficacy and reducing side effects (7). This approach is particularly useful in oncology, where nanoparticles loaded with anticancer drugs accumulate in tumor cells, improving treatment precision (8).

In electronics, **carbon nanotubes** and **quantum dots** represent significant breakthroughs. Carbon nanotubes, for example, are 100 times stronger than steel, but much lighter, making them invaluable for high-performance components in electronic devices and conductive materials. Quantum dots, semiconductor nanocrystals, enable the production of bright, colorful displays with low power consumption, used in TV screens and wearable devices. These advances in nanotechnology also open up prospects in energy technologies, notably for new-generation solar cells and more efficient batteries.

II. Artificial Intelligence and Digital Transformation

II.1 Using AI to optimize industrial processes

Artificial intelligence (AI) is transforming manufacturing processes in the chemical industry by optimizing resource consumption, improving safety, and increasing efficiency (9). Machine learning algorithms are being used to predict and prevent equipment failure through **predictive maintenance** systems (10). By analyzing sensor data in real time, AI can identify early warning signs of potential breakdowns, reducing production downtime and the costs associated with unscheduled repairs.

Automated control technologies, combined with AI, also enable chemical reaction parameters (temperature, pressure, concentration) to be optimized in real time, guaranteeing **efficient, energy-saving production** (11). In complex processes, AI adjusts dosages and the flow of raw materials to maximize yield and minimize waste, thus reducing the industry's environmental impact.

II.2 Big Data and molecular simulation in R&D

Big Data and AI-assisted molecular simulations are accelerating the research and development of new chemicals and materials (12). Massive databases containing information on chemical properties and molecular interactions enable researchers to **model complex chemical reactions** without the need for laboratory experiments, cutting costs and time. Molecular simulation software, used to design drugs or test new polymers, predicts the behavior of molecules under different conditions, offering an effective alternative to empirical testing.

In pharmaceutical chemistry, **machine learning** techniques analyze thousands of compounds to identify those with the potential to become active ingredients, accelerating the drug discovery process (13). At the same time, molecular simulations can be used to optimize formulations and improve the properties of materials for specific industrial applications (14). For example, chemical companies can use simulation to design polymers that are more resistant to extreme conditions, or more efficient catalysts for production reactions.

III. Sustainable Development and the Circular Economy

III.1 Chemical recycling and waste recovery

Chemical recycling represents a key innovation for reducing the environmental impact of waste plastics and other materials that are difficult to recycle mechanically (15). **Depolymerization** processes transform polymers into their original monomers, which can be reused to produce new materials without any degradation in quality. Pyrolysis, a thermal recycling method, breaks down plastic waste into basic hydrocarbons, which can be reintegrated into the production chain to manufacture virgin plastic or alternative fuels (16).

III.2 Example of the circular economy in the chemical industry

The circular economy in the chemical industry aims to reuse resources by maximizing their life cycle and minimizing waste (17). For example, some plastics manufacturers integrate production residues into their manufacturing processes, using waste as a secondary raw material. Thanks to materials chemistry, the food packaging industry is developing biodegradable films and bioplastics capable of degrading naturally, thus limiting environmental impact.

In addition, the chemical industry is working to **capture and reuse the CO_2** emitted during production processes (18). Recovered CO_2 can be transformed into methanol, which is then used as fuel, or incorporated into plastics and polymers, thereby helping to reduce the carbon footprint. Some chemical facilities also capture gaseous by-products for use in internal processes or for resale as commodity chemicals, reinforcing the idea of a circular economy where nothing is lost and everything is transformed.

These innovations are part of a **sustainable production** model, where companies not only reduce emissions and waste, but also seek to create value from by-products (19). The integration of the circular economy into industrial strategies reflects a profound change in the way the chemical industry sees its role, moving from a linear production approach to circular resource management.

References

1. Ratner, B., Hoffman, A., Schoen, F., & Lemons, J. E. (2013). *Biomaterials Science: An Introduction to Materials in Medicine*. Academic Press.
2. Otsuka, H., Nagano, S., & Higaki, Y. (2011). *Smart Polymers for Biomedical Applications*. Springer.
3. Hayes, S. A., & Seferis, J. C. (2001). "Self-healing composite materials". *Progress in Polymer Science*, 26(6), 933-957.
4. Ray, S. S., & Bousmina, M. (2005). *Biodegradable Polymers and Their Emerging Applications*. Taylor & Francis.
5. Ratner, M. A., & Ratner, D. (2003). *Nanotechnology: A Gentle Introduction to the Next Big Idea*. Prentice Hall.
6. Duncan, R. (2006). "Polymer therapeutics as nanomedicines: New perspectives". *Journal of Controlled Release*, 114(2), 91-108.
7. Ajayan, P. M., & Tour, J. M. (2007). "Nanotube super fibers". *Nature*, 447(7148), 1066-1068.
8. Smith, A. M., & Nie, S. (2010). "Semiconductor Nanocrystals: Structure, Properties, and Band Gap Engineering". *Journal of Physical Chemistry B*, 104(14), 2794-2801.
9. Lee, J., Bagheri, B., & Kao, H.-A. (2015). "A cyber-physical systems architecture for industry 4.0-based manufacturing systems". *Manufacturing Letters*, 3(1), 18-23.
10. Carvalho, T. P., et al. (2019). "A systematic literature review of machine learning methods applied to predictive maintenance". *Computers & Industrial Engineering*, 137, 106024.
11. Wan, J., et al. (2016). "Software-defined industrial Internet of Things in the context of industry 4.0". *IEEE Sensors Journal*, 16(20), 7373-7380.
12. Cummings, L., & DeVore, D. (2005). *Big Data Analytics for Chemists*. Springer.
13. Leach, A. R., & Gillet, V. J. (2007). *An Introduction to Chemoinformatics*. Springer.
14. Schneider, G., & Fechner, U. (2005). "Computer-based de novo design of drug-like molecules". *Nature Reviews Drug Discovery*, 4(8), 649-663.

15. Rahimi, A., & García, J. M. (2017). "Chemical recycling of waste plastics for new materials production". *Nature Reviews Chemistry*, 1(6), 1-11.

16. Christy, P., & Ramakrishna, S. (2016). *Anaerobic Digestion for Waste Treatment and Energy Production*. Elsevier.

17. Geissdoerfer, M., et al. (2017). "The Circular Economy-A new sustainability paradigm?". *Journal of Cleaner Production*, 143, 757-768.

18. Olah, G. A., Goeppert, A., & Prakash, G. K. S. (2006). *Beyond Oil and Gas: The Methanol Economy*. Wiley-VCH.

19. Ellen MacArthur Foundation. (2013). *Towards the Circular Economy*.

MIX
Papier aus verantwortungsvollen Quellen
Paper from responsible sources
FSC® C105338

Printed by Books on Demand GmbH, Norderstedt / Germany